Integration of PAM-IMAGING Chlorophyll Fluorometry
for Herbicide Sensitivity Estimations in Weed Research

Integration of PAM-IMAGING Chlorophyll Fluorometry for Herbicide Sensitivity Estimations in Weed Research

Dissertation to obtain the doctoral degree

of Agricultural Sciences (Dr. sc. Agr.)

Faculty of Agricultural Sciences

University of Hohenheim

Department of Weed Science (360b)

Institute of Phytomedicine

submitted by

Alexander Ingo Linn

from *Kirchheimbolanden*

2019

Bibliografische Information der Deutschen Nationalbibliothek

Die Deutsche Nationalbibliothek verzeichnet diese Publikation in der Deutschen Nationalbibliografie; detaillierte bibliographische Daten sind im Internet über http://dnb.d-nb.de abrufbar.

1. Aufl. - Göttingen: Cuvillier, 2020

Zugl.: Hohenheim, Univ., Diss., 2019

D100

© CUVILLIER VERLAG, Göttingen 2020

Nonnenstieg 8, 37075 Göttingen

Telefon: 0551-54724-0

Telefax: 0551-54724-21

www.cuvillier.de

ISBN 978-3-7369-7183-7

eISBN 978-3-7369-6183-8

Table of Contents

List of Tables

List of Figures

List of Abbreviations

ACCase	acetyl-Coa carboxylase
ALS	acetolactate synthase
CCD camera	charge-coupled device
CSV-file	comma-separated values
DAT	days after treatment
EXP	experiment
F_m	maximum fluorescence
F_o	ground fluorescence
F_v	variable fluorescence
F_v/F_m	maximum quantum efficiency of photosystem II
GPS	global positioning system
HRAC	herbicide resistance action committee
HSD	honestly significant difference
JPG	abbreviation of JPEG (Joint Photographic Experts Group)
L x W x H	length x width x width
LAN	local area network
LED	light-emitting diode
Li-Ion	lithium-ion
NDVI	normalised difference vegetation index
NIR	near-infrared
OS	operating system
PAM	pulse amplitude modulation
PRI	photochemical reflectance index
PS II	photosystem II
R	resistant
R?	slightly resistant
RR	moderate resistant
RRR	strongly resistant
S	sensitive
SE	standard error
STEME	*Stellaria media* L. Vill
TIFF	tagged image file format

Integration of PAM-IMAGING Chlorophyll Fluorometry for Herbicide Sensitivity Estimations in Weed Research

Summary

Since the introduction of synthetic herbicides, almost all mechanical weed control measures have been replaced by chemical control practices. Inevitably, the repeated weed control with herbicides, especially with the same mode of action, combined with the reduction of alternative indirect and direct weed control methods boosted the development of herbicide-resistant weeds. Fast in-field herbicide resistance detection is a key requirement for future herbicide resistance management, yet secure and rapid herbicide resistance tests are still missing. Chlorophyll fluorescence and chlorophyll fluorescence ratios can be very sensitive to changes in plant health status. Recent studies have demonstrated the potential of chlorophyll fluorescence measurements to evaluate the herbicide impact on crops and weeds. The fluorescence ratio F_v/F_m quantifies the maximum quantum efficiency of photosystem II (F_v/F_m) and indicates reliably plant health status. Hitherto, research on herbicide resistance detection via the F_v/F_m value have concentrated on *Alopecurus myosuroides* Huds. Investigations concerning herbicide resistance detection in other important weed species, both monocotyledonous as well as dicotyledonous, have not been carried out to a large extent. Therefore, more research is needed to investigate whether chlorophyll fluorescence, in particular F_v/F_m measurements, can be similarly useful in other agricultural crops and weeds. This work addresses the apparent gap in knowledge by four lines of research:

- Setup of a laboratory chlorophyll fluorescence agar test for the detection of herbicide resistance in *Apera spica-venti* L. Beauv;

- Development of a field-portable pulse amplitude modulated (PAM)-imaging chlorophyll fluorometer to determine plant stress in weeds and crops *in situ*;

- Field studies analyzing the F_v/F_m value for herbicide resistance detection in *Stellaria media* L. Vill and *Papaver rhoeas* L.;

- Design and testing of classifiers, in order to automatically assign individual herbicide-treated plants to the classes "resistant" or "susceptible" on the basis of their determined F_v/F_m value.

The outcome of the research has been published in three scientific papers:

1st paper: The development, application possibilities and limitations of a portable PAM-imaging chlorophyll fluorometer were reviewed. By measuring the fluorescence ratio F_v/F_m the device is capable to detect plant stress by herbicides, not only interacting directly with the photosystem II but also induced by herbicides acting on distant biological pathways. This stress can be determined within 1-5 days after treatment (DAT) before any damage becomes visible. Due to the compact design, the device is optimised to estimate herbicide sensitivity in crops and weeds in greenhouse and field studies. The device can be used for the detection of herbicide-resistant weed populations and single resistant specimens in combination with decision protocols. Furthermore, the herbicide sensitivity of crops can be estimated unbiased prior to visual assessments. It is concluded that the PAM-imaging sensor is suitable as an expert tool for decision support in integrated weed management.

2nd paper: In order to validate the possibility of accelerating screenings by a F_v/F_m survey, an agar test on herbicide resistance in *A. spica-venti* was investigated. Sensitive and resistant *A. spica-venti* plants were cultivated in herbicide treated agar. The plant stress was quantified by the F_v/F_m value with the PAM-imaging chlorophyll fluorometer. Though plant stress increased with increasing herbicide concentrations in all experiments, the results of the agar test on herbicide sensitivity were highly variable and did not represent results from standard greenhouse pot trials. It is assumed that the variability in the results is due to the increased herbicide concentrations used for inducing the rapid plant response.

3rd paper: The sensor capabilities for herbicide resistance detection in dicotyledonous weeds have been investigated in the field. Herbicide-resistant and sensitive *S. media* and *P. rhoeas* plants were treated with acetolactate-synthase inhibitors. The plant stress was quantified by the determination of the F_v/F_m value with the portable PAM-imaging chlorophyll fluorometer. Discriminant maximum likelihood classifiers were created and tested with independent data sets, assigning single plants based on the measured F_v/F_m values to the classes "resistant" or "susceptible". The F_v/F_m values of sensitive *P. rhoeas* and *S. media* plants decreased within 3 DAT by 28–43%. The F_v/F_m values of the resistant plants were 20% higher than those of the sensitive plants in all herbicide treatments. The classifiers separated sensitive and resistant plants within 3 DAT with accuracies of 62% to 100%. Therefore, an estimation of herbicide resistance can be performed within 3 days after herbicide treatment.

In summary, these papers demonstrate that a mobile PAM-imaging chlorophyll fluorometer can perform fast estimations of herbicide stress in laboratory and field conditions. By employing this PAM device, it is shown that the detection of herbicide action by F_v/F_m also works for dicotyledonous species. Moreover, it was possible to identify individual herbicide-resistant plants based on its value of F_v/F_m. Hence, PAM fluorometry can play a significant role in a decision support system in integrated weed management.

Zusammenfassung

Synthetische Herbizide haben seit deren Einführung fast alle mechanischen Unkrautbekämpfungsmaßnahmen verdrängt. Jedoch fördert der wiederholte Einsatz von Herbiziden - oftmals mit gleicher Wirkungsweise -, kombiniert mit einer Reduzierung indirekter und anderer direkter Bekämpfungsmaßnahmen, zwangsläufig die Selektion herbizidresistenter Unkrautpopulationen. Eine frühe Erkennung dieser resistenten Unkrautpopulationen ist ein zentraler Baustein des Resistenzmanagements. Eine Möglichkeit Veränderungen im Gesundheitszustand von Pflanzen schnell zu erkennen, bieten Chlorophyllfluoreszenzmessungen, da die Chlorophyllfluoreszenz sowie Chlorophyllfluoreszenzquotienten sehr empfindlich auf diese Veränderungen reagieren können. Jüngere Studien zeigten bereits das Potential von Chlorophyllfluoreszenzmessungen zur Bewertung, der durch Herbizide induzierten Veränderungen in Kulturpflanzen und Unkräutern. Der Fluoreszenzquotient F_v/F_m quantifiziert die maximale Quantenausbeute des Photosystems II (F_v/F_m) und ermittelt somit zuverlässig Pflanzenstress. Bisher konzentrierte sich die Forschung hauptsächlich auf *Alopecurus myosuroides* Huds., um herbizidresistente Unkräuter anhand des F_v/F_m-Wertes nachzuweisen. Umfangreiche Studien zur Detektion weiterer monokotyler und dikotyler herbizidresistenter Unkrautpopulationen unter Verwendung des F_v/F_m-Wertes sind bislang rar. Um zu evaluieren, ob sich der F_v/F_m Wert eignet Herbizidresistenz in weiteren Kulturpflanzen und Unkräutern zu erkennen, wurden in dieser Arbeit folgende Forschungsschwerpunkte bearbeitet:

- Entwicklung eines chlorophyllfluoreszenzbasierenden Agartests im Labor zum Nachweis der Herbizidresistenz in *Apera spica-venti* L. Beauv.;
- Entwicklung eines tragbaren, bildgebenden Puls-Amplituden-Modulations (PAM)-Chlorophyllfluorometers zur Erkennung des Pflanzenstresses in Unkräutern und Kulturpflanzen *in situ*;
- Feldstudien zur Bewertung des F_v/F_m-Wertes zum Nachweis der Herbizidresistenz in *Stellaria media* L. Vill. und *Papaver rhoeas* L.;
- Erstellung und Prüfung von Klassifikatoren, um einzelne herbizidbehandelte Pflanzen anhand ihres F_v/F_m-Wertes automatisch den Klassen „resistent" oder „sensitiv" zuzuordnen.

Die Ergebnisse wurden in drei wissenschaftlichen Veröffentlichungen zusammengefasst und diskutiert:

<u>1. Artikel</u>: Es wurde die Eignung eines tragbaren PAM-Chlorophyllfluorometers zur Ermittlung von herbizidinduziertem Pflanzenstress, dessen Anwendungsbereiche und systembedingte Limitierungen untersucht. Aufgrund der kompakten Bauweise des PAM-Fluorometers eignet es sich hervorragend für Studien zur Quantifizierung der Herbizidempfindlichkeit von Pflanzen im Gewächshaus und Feld. Durch die Messung des Chlorophyllfluoreszenzquotienten F_v/F_m ist das tragbare, bildgebende PAM-Fluorometer in der Lage herbizidinduzierten Pflanzenstress zu identifizieren. Dies gilt nicht nur für Herbizide, die direkt mit dem Photosystem II interagieren, sondern auch für Herbizide, deren Wirkort außerhalb des Photosystems II liegt. Dieser herbizidinduzierte Pflanzenstress kann innerhalb von 1-5 Tagen nach der Herbizidbehandlung (DAT) ermittelt werden. Zu diesem Zeitpunkt sind visuell noch keine Schäden an der Pflanze erkennbar. In Kombination mit Entscheidungsprotokollen kann dieses PAM-Fluorometer sowohl zum Nachweis herbizidresistenter Unkrautpopulationen als auch einzelner resistenter Pflanzen eingesetzt werden. Darüber hinaus ist eine objektive Einschätzung der Herbizidsensitivität in Kulturpflanzen noch vor einer visuellen Bonitur möglich. Zusammenfassend ist das bildgebende PAM-Chlorophyllfluorometer als Werkzeug zur Einschätzung der Herbizidsensitivität von Kulturpflanzen sowie zur Detektion von Herbizidresitenz in Unkräutern geeignet.

<u>2. Artikel</u>: Um die Erkennung herbizideresistenter Unkräuter mit Hilfe von F_v/F_m-Messungen zu beschleunigen, wurde ein Agartest zur Herbizidresistenzdetektion in *A. spica-venti* untersucht. Herbizidsensitive und -resistente *A. spica-venti* Pflanzen wurden in herbizidhaltigem Agar etabliert. Der Pflanzenstress wurde mit einem PAM-Chlorophyllfluorometer anhand des F_v/F_m-Wertes quantifiziert. Obwohl in allen Experimenten der Pflanzenstress mit zunehmender Herbizidkonzentration zunahm, waren die Ergebnisse des Agartests sehr variabel und spiegelten die erhobenen Ergebnisse aus herkömmlichen Resistenztests im Gewächshaus nicht wieder. Es wird davon ausgegangen, dass die Variabilität der Ergebnisse auf die im Agartest verwendeten, erhöhten Herbizidkonzentrationen zurückzuführen ist.

<u>3. Artikel</u>: In Feldversuchen wurde die Möglichkeit für den Nachweis herbizidresistenter dikotyler Unkräuter unter Zuhilfenahme des tragbaren PAM-Chlorophyllfluorometers untersucht. Herbizidresistente und -sensitive *S. media* und *P. rhoeas* Pflanzen wurden mit Acetolactat-Synthase-Hemmern behandelt. Der Pflanzenstress wurde anhand des F_v/F_m-Wertes mit dem tragbaren PAM-Chlorophyllfluorometer quantifiziert. Es wurden, auf den gemessensen F_v/F_m-Werten basierende, diskriminierende Maximalwahrscheinlichkeits-Klassifikatoren erstellt und mit unabhängigen Datensätzen getestet. Die Klassifikatoren

ordneten einzelne Pflanzen den Klassen „resistent" oder „sensitiv" zu. Die F_v/F_m-Werte von sensitiven *P. rhoeas* und *S. media* Pflanzen sanken innerhalb von 3 DAT um 28-43%. Die F_v/F_m-Werte der resistenten Pflanzen waren bei allen Herbizidbehandlungen 20% höher als die der sensitiven Pflanzen. Die Klassifikatoren ordneten sensitive und resistente Pflanzen innerhalb von 3 DAT mit Genauigkeiten von 62% bis 100% ein. Dies ermöglicht eine Einschätzung zur Herbizidresistenz in *S. media* und *P. rhoeas* binnen 3 DAT.

Zusammenfassend zeigen diese Arbeiten, dass die PAM-Chlorophyllfluorometrie zur Bestimmung des Pflanzengesundheitszustandes innerhalb weniger Tage nach der Herbizidapplikation im Labor sowie im Feld geeignet ist. Unter Zuhilfenahme des mobilen PAM-Fluorometers konnten herbizidresistente dikotyle Unkräuter durch Bestimmung des F_v/F_m-Wertes identifiziert werden. Darüber hinaus war eine automatisierte Erkennung einzelner herbizidresistenter Pflanzen anhand des F_v/F_m-Wertes möglich. Damit kann die PAM-Chlorophyllfluoreszenzmessung einen wertvollen Beitrag zum Resitenzmanagement leisten.

Chapter I

General Introduction

1 General Introduction

Herbicide-resistant weeds constitute one of the most serious problems in agriculture. In the current agricultural practice, almost all mechanical weed control measures have been replaced by herbicides (Heap, 2014). The strong and repeated selection pressure by herbicides forced the development of herbicide-resistant weed populations. Since the first documentation of herbicide-resistant weed plants in 1957 (Hilton, 1957; Switzer, 1957), the number of herbicide-resistant weed species and populations, and the affected herbicide modes of action increased steadily (Heap, 2019). With the rise of herbicide-resistant weed populations, herbicide resistance management became a fundamental part of crop management. An integral part of herbicide resistance management is the safe and rapid testing for herbicide resistance.

The pulse amplitude modulated (PAM)-imaging chlorophyll fluorometry is an analytical tool to study photosynthesis. Chlorophyll fluorescence parameters and ratios can be very sensitive indicators for changes in the plant health status. The maximum quantum efficiency of photosystem II (F_v/F_m) is a chlorophyll fluorescence ratio that represents the performance of photosystem II photochemistry (Baker, 2008). After the determination of the ground fluorescence (F_o) and the maximum fluorescence (F_m) of dark acclimated plants, the F_v/F_m value can be derived ($F_v/F_m = (F_m-F_o)/F_m$). In most higher plants, the F_v/F_m value ranges from 0.78 and 0.84 in healthy conditions, while lower F_v/F_m values indicate plant stress (Stirbet & Govindjee, 2011). The F_v/F_m value quantifies plant stress, independent of the stress origin (Rosenqvist & van Kooten, 2003). Therefore, the F_v/F_m has proven to be a reliable parameter for plant stress identification.

The potential of PAM-imaging chlorophyll fluorometry for herbicide sensitivity assessments using the F_v/F_m value in weeds and crops has already been demonstrated (Kaiser *et al.*, 2013; Wang *et al.*, 2016; Wang *et al.*, 2018; Weber *et al.*, 2017; Li *et al.*, 2017). By employing the PAM-imaging sensor, herbicide-induced plant stress in crops can be quantified by determining the F_v/F_m value and thereby deriving statements on the crop sensitivity to the herbicides as well as variety-related herbicide sensitivities (Weber *et al.*, 2017). Likewise, the F_v/F_m value can provide information on the herbicide resistance status of weed populations after herbicide application even before visible symptoms of herbicide damage occur (Kaiser *et al.*, 2013; Wang *et al.*, 2018; Wang *et al.*, 2016). Here, the research focus was on *Alopecurus myosuroides* Huds. Since the potential for the detection of herbicide-resistant *A. myosuroides* has been demonstrated, it should be investigated whether conclusions about the herbicide-resistance status can also be drawn in other weed species, monocotyledonous as well as dicotyledonous. This would significantly expand the range of applications for the PAM-imaging chlorophyll fluorometry and enhance the system's versatility. A mobile PAM-imaging chlorophyll fluorometer for field use was introduced for the use of the PAM-imaging chlorophyll fluorometry outside laboratories in field trials (Wang *et al.*, 2016). Since that the fluorometer was continuously advanced. Due to the many applications of PAM-imaging chlorophyll fluorometry with respect to the detection of herbicide-induced plant stress, a detailed description of the system, the discussion of possible areas of application, and limitations for its use in arable farming are necessary.

1.1 Objectives

The aims of this research work were: First, the adjustment of a mobile PAM-imaging chlorophyll fluorometer to identify the induced stress in agricultural plants by measuring the F_v/F_m value. Second, to develop a modification of a pre-existing laboratory agar test for the detection of herbicide resistance in *Apera spica-venti* (L.) Beauv by the determination of the F_v/F_m value after the herbicide treatment. Third, to investigate the employment of F_v/F_m value for the identification of herbicide-resistant *Papaver rhoeas* L. and *Stellaria media* L. Vill populations in-field, and, further, to create an autonomous assignment of single herbicide treated *P. rhoeas* and *S. media* plants based on their F_v/F_m value to the classes "susceptible" or "resistant".

1.2 Structure of the Dissertation

The dissertation is presented in a cumulative thesis of three articles. At the current state, one article is submitted and under review in a peer-reviewed journal. The other two articles are published in peer-reviewed journals. The publications are presented in a unified format and reference style.

The first article entitled "Development and applications of a field imaging chlorophyll fluorometer to measure stress in agricultural plants" is submitted to the Journal "Precision Agriculture". This review describes a mobile PAM-imaging chlorophyll fluorometer, including the special physical properties of the sensor and gives examples of applications.

The second article entitled "Detecting herbicide-resistant *Apera spica-venti* with a chlorophyll fluorescence agar test" is published in the Journal "Plant Soil and Environment". It describes the approach of adopting a chlorophyll fluorescence agar test for herbicide resistance detection in *A. spica-venti*.

The third article entitled "In-field classification of herbicide-resistant *Papaver rhoeas* and *Stellaria media* using an imaging sensor of the maximum quantum efficiency of photosystem II" is published in the Journal "Weed Research". It describes the introduction of PAM-imaging chlorophyll fluorometry for the detection of herbicide-resistant *S. media* and *P. rhoeas* in the field. In addition, a classifier is presented that assigns single plants to the classes "susceptible" and "resistant" based on their F_v/F_m value after herbicide treatment.

Besides the presented publications in this thesis, three contributions to international conferences were accepted:

- A. Menegat, B. Sievernich, A. Linn, R. Mink & R. Gerhards (2016): Development of a standardised test system for detection of resistance against per-emergence herbicides. *Proceedings 7th International Weed Science Congress.* "Weed science and management to feed the planet", Prague, Czech Republic. ISBN 978-80-213-2648-4

- A. Linn, R. Mink, G. Peteinatos & R. Gerhards (2018): Herbicide efficacy estimation of ALS-inhibitors in *Stellaria media* L. and *Papaver rhoeas* L. *Proceedings 18th European Weed Science Society Symposium.* "New approaches for smarter weed management", Ljubljana, Slovenia. ISBN 978-961-6998-21-5

- P. Kosnarova, K. Hamouzova, A. Linn, P. Hamouz & J. Soukup (2018): *Apera spica-venti* biotype from the Czech Republic resistant to three herbicide modes of action. *Proceedings 18th European Weed Science Society Symposium.* "New approaches for smarter weed management", Ljubljana, Slovenia. ISBN 978-961-6998-21-5

Chapter II

Publications

2 Publications

2.1 Features and Applications of a Field Imaging Chlorophyll Fluorometer to Measure Stress in Agricultural Plants

Alexander I. Linn[1]*, Alexander K. Zeller[1], Erhard E. Pfündel[2], Roland Gerhards[1]

[1]Institute of Phytomedicine, Department of Weed Science, University of Hohenheim, Stuttgart, Germany.

[2]Heinz Walz GmbH, Effeltrich, Germany.

Submitted to: Precision Agriculture; ISSN 1385-2256

2.1.1 Abstract

Most non-destructive methods for plant stress detection do not measure the primary stress response but reactions of processes downstream of primary events. For instance, the chlorophyll fluorescence ratio F_v/F_m, which indicates the maximum quantum yield of photosystem II, can be employed to monitor stress originating elsewhere in the plant cell. This article reviews the development of a sensor to quantify herbicide stress in agricultural plants for field applications by the F_v/F_m parameter. This dedicated sensor is highly mobile and measures images of pulse amplitude modulated (PAM) chlorophyll fluorescence. Special physical properties of the sensor are reported and the range of its field applications is defined. In addition, detection of herbicide-resistant weeds by employing an F_v/F_m-based classifier is described. Examples are presented to illustrate the potential of the sensor as a general indicator of stress on crops and weeds. It is concluded that stress detection by the F_v/F_m parameter is suitable as an expert tool for decision making in crop management.

Keywords chlorophyll fluorescence; classifier; decision support system; plant stress; pulse amplitude modulation fluorometer

2.1.2 Introduction

With the need to reduce pesticide use in agriculture while maintaining high yields and food quality, the precise monitoring of crop physiological status gains increasing importance. In the past four decades, several optical sensor systems have been developed to support farmers and consultants by identifying crop stress response to fungal infections and diseases, weed competition and pests (Peteinatos *et al.* 2016). Sensors have also been used to detect phytotoxic side effects of pesticides in crops (Weber *et al.* 2016). Pests, fungal diseases and weeds can be controlled with lower pesticide rates, when they are detected in an early stage of development. However, selection of the best pesticide is very difficult due to the worldwide emergence of resistant populations of weeds, fungi and pests to pesticides. These challenges call for new sensor systems allowing rapid and early plant stress detection under field conditions.

Spectrometers measuring the reflected part of radiation determine the physiological status of the plants non-intrusively (Pascua *et al.*, 2019). This type of sensor employs established relationships between reflected radiation and biochemical and structural properties of plants (Knipling, 1970; Xue & Su, 2017). Generally, reflectance spectra of non-stressed plants are characterized by low values in the visible range due to high absorbance from plant pigments such as chlorophylls and xanthophylls, and high reflectance in the near infra-red (NIR) range where these pigments do not absorb (Blackburn, 2006). In stressed plants, often decreased pigment concentrations increase visible reflectance.

For plant stress detection by reflectance data, indices have been introduced which combine specific ranges of reflected visible light and employ NIR wavelengths as reference (Govender *et al.*, 2009; Peteinatos *et al.*, 2016). The most known index is the normalized difference vegetation index (NDVI) (Rouse *et al.*, 1973). Indices such as the photochemical reflectance index (PRI) positively correlate with photosynthetic efficiency (Gamon *et al.*, 1992).

Since standard spectrometers do not provide spatial information, the stress affecting only small parts of the observed area may not be detected. To detect disease spots in an apparently healthy region, imaging spectrometers such as multi-spectral cameras, hyperspectral cameras and line scanners have been developed (Bauriegel & Herppich 2014; Lowe *et al.*, 2017).

Spatial information is also obtained by thermal imaging sensors (Vadivambal & Jayas, 2011). They measure infra-red radiation by special detectors and display the originally greyscale images in pseudo-color. The spatial and spectral resolution of thermal cameras is high enough to detect small spots of higher or lower leaf temperature. Higher temperatures of living plant tissue are often associated with lower transpiration and might indicate a fungal infestation (Lindenthal *et al.*, 2005). In necrotic leaf tissue, leaf temperature is often lower. A suitable parameter to separate healthy from diseased crop canopies is the maximum temperature variation (Oerke & Steiner, 2010).

The present paper focusses on another imaging sensor, which is based on measurements of chlorophyll fluorescence. Chlorophyll fluorescence, together with heat, represent losses of light energy absorbed by the photosynthetic apparatus (Baker, 2008). In predarkened leaves, the emission of fluorescence ranges between 2-10% of the absorbed light depending on the state of photosystem II (PS II, Trissl *et al.*, 1993). A relationship between the intensity of chlorophyll fluorescence and the state of photosynthesis has already been observed by Kautsky & Hirsch (1931). In such experiments, sudden light exposure of a leaf first increases chlorophyll fluorescence within a second which then drops off slowly within a few minutes. The initial rise is believed to reflect the transition from open to closed PS II reaction centers, that is, from a photochemically competent state to a state in which light energy cannot be utilized for photochemistry (Schreiber *et al.*, 1995).

In PS II, light-driven charge separation by the reaction center initiates photochemistry. This reaction ensues electron transfer via pheophytin to Q_A, a plastoquinone molecule bound to the PS II reaction center (Loll *et al.*, 2005). In the presence of reduced Q_A^-, the PS II reaction center is closed because it cannot use excitation energy for stable charge separation (Hankamer & Barber, 1997). Usually, the Q_A is quickly reoxidized by electron transfer to Q_B of the reaction center, which after a second reduction is exchanged by an oxidized Q_B of the photosynthetic electron transport chain. Very strong (saturating) light intensities result in reduction of this Q_B pool so that Q_A^- reoxidation stops and PS II reaction centers remain closed (Schreiber *et al.*, 1995).

The competition for excitation energy between fluorescence emission and photochemistry explains why minimum fluorescence (F_o) is observed in the dark, that is, in the presence of open PS II reaction centers, but maximum fluorescence (F_m) when these reaction centers are closed. Already Kitajima & Butler (1975) have derived that the fluorescence ratio F_v/F_m (where $F_v=F_m-F_o$) is related to the maximum photochemical efficiency of PS II. Until today, the F_v/F_m parameter plays an important role in plant stress research (Schreiber *et al.*, 1995; Weber *et al.*, 2016; Li *et al.*, 2017; Poudyal *et al.*, 2019).

Using chlorophyll fluorescence analysis to quantify plant stress caused by herbicides is well established (Macinnis-Ng & Ralph, 2003; Follak & Hurle, 2004; Bigot *et al.*, 2007; Jin *et al.*, 2010). This application is of general interest because herbicide treatment is the most effective and most frequently used weed control strategy in conventional agriculture (Heap, 2014).

An ideal herbicide would selectively kill the weeds but not harm the crop. In reality, even very selective herbicides can damage the crop (Gerhards *et al.*, 2011). This damage may increase when the selectivity of a herbicide is compromised by unfavorable weather conditions, inappropriate timing of the application, wrong rates, and mixtures of active ingredients as well as the addition of adjuvants can reduce the selectivity (Salzman & Renner, 1992). Indeed, insufficient selectivity of herbicides have been reported to cause crop biomass reduction and yield losses (Wilson, 1999; Wilson *et al.*, 2002; Belfry *et al.*, 2015). The selectivity of a herbicide also depends on the cultivar of the crop plant: for instance, elevated capability to detoxify the active ingredient can reduce the noxiousness of a herbicide for the crop (Smith & Wilkinson, 1974). Clearly, quantifying herbicide stress on both weed and crop is important to assess the usefulness of an herbicide.

Initially, measurements of F_v/F_m were confined to specialized laboratories but the commercial availability of pulse amplitude modulated (PAM) fluorometers has made analyses of photosynthesis by measuring F_v/F_m and other fluorescence ratio parameters a common analytical tool. To employ PAM fluorometry for detection of plant stress under field conditions, hardware and software modifications of an existing PAM imaging device have been changed. These developments and the typical measurement protocol are summarized in the present paper.

This review further highlights that the PAM-imaging sensor introduced here can provide in-field estimation on herbicide sensitivity already a few days after treatment (DAT) with herbicide, that is, before any damage becomes visible. We also cover limitations of the sensor system and potential further applications for chlorophyll fluorescence imaging in crop science including the use of a classifier to differentiate between stressed and non-stressed plants based on sensor data.

2.1.3 Methods

Hardware of the PAM-sensor

The PAM-imaging sensor is a mobile MINI Version of the IMAGING-PAM M-Series Chlorophyll Fluorometer (Heinz Walz GmbH, Effeltrich, Germany). The device consists of a waterproof tablet computer, a multi control unit and a measuring unit (labels 1 to 3 in Fig. 1a), where tablet computer and multi control unit are mounted on an aluminum frame with carrying handle. The measuring unit consists of a 1/3" CCD camera (640x480 pixels) with K7-MIN prime lens (F1.4/f=16 mm) and an LED array arranged around the camera lens. The LED array consists of 12 blue (470 nm) LEDs (Fig. 1b) that deliver both the pulse-modulated excitation light and the saturation pulse which drives the fluorescence rise from F_o to F_m. The LEDs are equipped with a bandpass filter (BG39) to prevent their weak long-wavelength emission from reaching the detector. A long pass filter is mounted in front of the camera transmitting only fluorescence radiation and excluding wavelengths below 640 nm. The captured fluorescence is digitized in the camera which is connected to the tablet computer via Ethernet. The multi control unit contains a microcontroller and a rechargeable Li-Ion battery (14.4 V / 6 Ah) powering camera and LEDs.

To dark-acclimate plants for F_o measurements, blackout boxes (LxWxH = 0.19 m x0.19 x 0.8 m) were designed. The bottom of the box forms an aluminum frame whose center is placed above the target plant (Fig. 1c). The frame features spikes for a firm fit on ground and a seat for the upper parts of the box. The box walls are formed by a deep-drawn plastic piece having the shape of a truncated pyramid. The top of the box can be sealed against light by a slider and it also functions as a socket for the measuring unit (Fig. 1d). A Nylon net is mounted 3.5 cm above ground to keep larger plants in the focus level of the camera. The measuring area is 24 x 32 mm². Dark acclimation of the plants is done in the blackout boxes with closed slider for at least 20 min. After positioning the measuring unit on the box, the slider is opened, so that fluorescence can be measured with the still dark-acclimated plant.

Fig. 1 Overview of the hardware and accessories of the new PAM-imaging system. **a** Tablet computer (1), control unit (2) and measuring unit (3). **b** Measuring head with blue light-emitting diodes and central camera lens. **c** Aluminum frame with target plant. **d** Blackout box

PAM fluorescence and saturation pulse analysis

The PAM-imaging sensor measures only the fluorescence excited by µs flashes of probing light. The effective intensity of these measuring flashes is very weak so that virtually all PS II reaction centers remain open. Under such conditions, F_o fluorescence is measured. To close the PS II reaction centers and, thus, to elicit F_m fluorescence, the PAM-imaging sensor applies a high intensity pulse of several hundreds of ms length and of 8200 µmol m^{-2} s^{-1} photosynthetic photon flux density (wavelength 470 nm). The fluorescence ratio of F_v/F_m is a measure for the maximum quantum yield of primary PS II photochemistry. In healthy condition, the value for most higher plants is between 0.78 and 0.84 (Stirbet & Govindjee, 2011). A lower value indicates decreased performance of PS II photochemistry. This is called photoinhibition, which is often observed in plants under stress conditions (Abbaspoor & Streibig, 2005).

Software

The software of the PAM-imaging sensor is based on the ImagingGigE software for IMAGING-PAM fluorometers (Heinz Walz GmbH). The software acquires images with 640 x 480 pixels for F_o and F_m level fluorescence of the same sample. From these images, an image of F_v/F_m is created by pixelwise calculations. The originally greyscale image of F_v/F_m is converted into colored information using a false color scale. The false color scale results from stepwise increasing the angular position around the axis of an HSL (hue, saturation, lightness) color space cylinder where the S and the L are kept constant. Figure 2 shows false color images of F_v/F_m for a stressed and a non-stressed plant of *Stellaria media* (L.) Vill. together with the false color scale. The bluish coloration of the non-stressed plant indicates an F_v/F_m of about 0.8 (Fig. 2b) which was decreased to around 0.4 in the stressed plant (Fig. 2a).

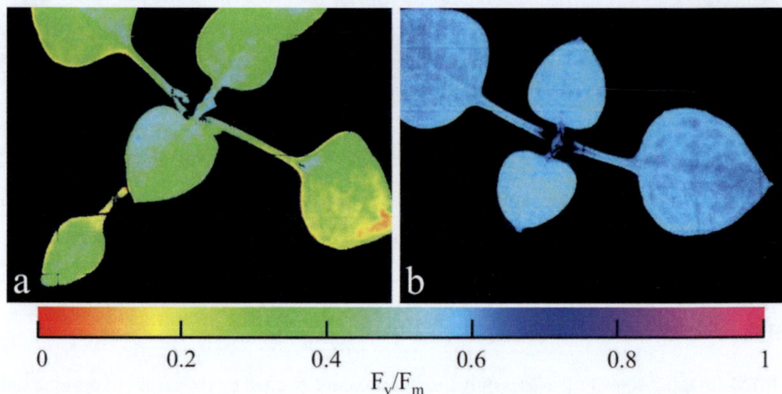

Fig. 2 Images of the F_v/F_m parameter of stressed (**a**) and non-stressed (**b**) *Stellaria media* (L.) Vill. The false color scale results from stepwise increasing the angular position around the axis of an HSL (hue, saturation, lightness) color space cylinder where the S and the L are kept constant. The false-colour bar for F_v/F_m is shown below the images

The software permits export of alphanumeric data as plain text (CSV file) and pictures of F_o, F_m and F_v/F_m as TIFF and JPG images. Data handling follows good laboratory practice and does not allow changes of the raw data. Post-processing of data is possible on computers with Windows OS in the offline mode of the software. For post-processing, the area of interest can be assigned manually and several areas can be selected at a time. For each defined area of interest, the arithmetic mean for F_v/F_m value is calculated automatically by the software.

Depending on the size of the sample, an area may represent a whole plant, various leaves or only a leaf part. Post-processing also allows the setting of global upper and lower thresholds for pixel values of F_v/F_m. Because background fluorescence is often lower than F_o fluorescence of the target plant, proper setting of the threshold blanks out fluorescing background and, thus, excludes it from F_v/F_m calculations. For field use, data management of the ImagingGigE software was modified to permit storage of consecutive F_v/F_m measurements in a single file..

The original software disregarded areas exhibiting F_v/F_m values close to zero. This software property was designed to distinguish photosynthetic from non-photosynthetic matter. However, strong herbicide action can diminish the F_v/F_m of parts of leaves close to zero with the result that these areas are ignored and calculations of the leaf F_v/F_m become biased towards the less affected leaf regions. In case of grass leaves, this potential difficulty was overcome by introducing the ability of pattern recognition which identifies the leaf based on shape resulting in evaluation of the entire leaf including areas exhibiting very low F_v/F_m. For broad-leaved weeds, the database for pattern recognition is pending..

To increase the traceability of measurements, a global positioning system (GPS) module was implemented writing directly the coordinates of the investigated area into the data file. In addition, the camera of the tablet can be used for barcode scanning of a plant sample to assign the sample directly to the measurement. After finishing the measurements, data files can be uploaded automatically to a cloud server via Wi-Fi or LAN connection to the internet. Upload of data to a cloud server was implemented solely with the objective of documentation and sharing of original data. A cloud-based automated post-processing routine is currently not available..

2.1.4 Applications

This section gives an overview on experiments performed with the PAM-imaging sensor. The described experiments are summarized in Table 1. Further experimental descriptions can be found in the respective references.

Herbicide induced stress in weeds

Monitoring of the herbicide efficacy on weeds plays a major role in crop production. The earlier insufficient herbicide action is detected, the sooner additional weed control measures can be applied. In the case of herbicide resistant weeds, timely countermeasures following their early detection can reduce the risk of growing resistant populations. Therefore, a strategy of estimating the herbicide efficacy, including the detection of herbicide resistant weeds, was developed in three main steps.

In step 1, weed sensitivity was estimated on the population level by comparing the mean F_v/F_m values of herbicide treated and untreated *Alopecurus myosuroides* Huds. Plants (EXP1, EXP2, EXP3, Table 1). In the second step, the F_v/F_m values of susceptible and unsusceptible single plants were separated *post-hoc* to investigate the differences in the F_v/F_m values of these two sample types (EXP4, Table 1). In step 3, a classifier was created to assign single plants to the classes "sensitive" and "resistant" based on their F_v/F_m value. Ideally, this classifier should identify herbicide resistance without untreated reference plants (EXP5, Table 1). These three steps are explained in more detail in the three subsequent sections.

Herbicide sensitivity estimation on population level

Wang *et al.* (2016) described the first estimations on herbicide-induced stress in *A. myosuroides* with the PAM-imaging sensor under field conditions (EXP1, Table 1). This first field experiment formed the basis of later studies with this sensor on herbicide resistance. The authors examine the discrimination of sensitive *A. myosuroides* plants from untreated plants after treatment with acetolactate synthase (ALS) inhibitors and acetyl-CoA carboxylase (ACCase) inhibitors in a field experiment. The differences in the F_v/F_m values of the untreated and sprayed plants were evaluated using Tukey's honestly significant difference (HSD) test per day ($P < 0.05$) with the result that a discrimination of untreated and treated plants was possible for ALS- and ACCase-inhibiting herbicides from 5 DAT on.

Tab. 1 Overview on the described experiments. HRAC – herbicide resistance action committee; EXP – experiment; RCD – randomized complete design; RCBD – randomized complete block design; No. – Number of populations or cultivars

EXP	Target species	Sample size	No.	Stressor (for herbicides: HRAC-group)	Design	Location	Section	Reference
1	*Alopecurus myosuroides* Huds.	40 plants per plot	1	Herbicide (A, B)	RCBD; 4 blocks	field	Herbicide induced stress in weeds	Wang *et al.*, 2016
2	*Alopecurus myosuroides* Huds.	40 plants per plot	8	Herbicide (A, B)	RCBD; 4 blocks	field	Herbicide induced stress in weeds	Wang *et al.*, 2016
3	*Alopecurus myosuroides* Huds.	40 plants per plot	42	Herbicide (A, B)	RCBD; 4 blocks	field	Herbicide induced stress in weeds	Wang *et al.*, 2016
4	*Alopecurus myosuroides* Huds.	40 plants per plot	4	Herbicide (A, B, C)	RCBD; 4 blocks	field	Herbicide induced stress in weeds	Wang *et al.*, 2018
5	*Stellaria media* L. Vill	10 plants per treatment	2	Herbicide (B)	RCD	greenhouse	Herbicide induced stress in weeds	unpublished
6	*Beta vulgaris* spp. *Vulgaris*	1 plant per plot	9	Herbicide (C, K, O)	RCBD; 4 blocks	greenhouse	Herbicide induced stress in crops	Weber *et al.*, 2016
	Glycine max (L.) Merr.		7					
7	*Glycine max* (L.) Merr.	10 plants per plot	1	Herbicide (C, F, K)	RCBD; 4 blocks	field	Herbicide induced stress in crops	Li *et al.*, 2017
8	*Zea mays* L.	5 plants per treatment	3	Herbicide (B, C, F, K)	RCBD; 4 blocks	greenhouse	Herbicide induced stress in crops	unpublished
9	*Zea mays* L.	7 plants per plot	1	Herbicide (B, C, F, K)	RCBD; 4 blocks	field	Herbicide induced stress in crops	unpublished
10	*Triticum aestivum* L.	4 per block	1	Pathogen (*Blumeria graminis* f. sp. *Tritici*) Competition (*Sinapis arvensis* L.)	RCBD; 3 blocks	greenhouse	Weed competition and fungi infection	unpublished

For the further development of the PAM-imaging system, two large-scale field trials were carried out. These trials focused on the distinction between herbicide-resistant and sensitive weed populations (Wang *et al.*, 2016). The first set of experiments was carried out at eight locations in northern and southern Germany with the fields' native *A. myosuroides* populations (EXP2, Table 1). For the second experiment, 42 *A. myosuroides* populations with known resistance status were sown into a winter wheat field at one site (EXP3, Table 1). In both experiments, untreated controls per population were included and each population was treated with ALS and ACCase inhibiting herbicides. Measurements of F_v/F_m values were performed at 5 DAT. The F_v/F_m value-based classification of the populations (arithmetic mean of 40 plants plot^{-1} Tukey's HSD test, P < 0.05) was clearly supported by the proportion of dead and alive plants 21 DAT (Table 2)

Tab. 2 Confusion matrix of visually and fluorometrically assessed herbicide resistance of *Alopecurus myosuroides* Huds. populations after treatment with acetolactate synthase and acetyl-coA carboxylase inhibitors, respectively. Plots of 40 plants were evaluated. Visual and fluorometric assessment was 21 and 5 days after treatment, respectively. Visual assessment is based on the proportion of surviving plants. The PAM-imaging sensor was used for fluorometric assessment. S = sensitive, 100-81 % dead; R = resistant, 80-0% dead. In the visual assessment R represents the summarized groups of R? + RR + RRR. R? = slightly resistant, 80-73% dead + RR = moderate resistant, 72-37% dead + RRR = strongly resistant, 36-0% dead. (Wang *et al.*, 2016; see text section *Herbicide sensitivity estimation on population level*)

		Visual assessment	
		S	R
PAM-imaging sensor	S	27	11
	R	0	58

The findings of these experiments (EXP1, EXP2, EXP3) (Wang *et al.*, 2016) stated that the F_v/F_m value obtained from 40 untreated and 40 herbicide-treated plants gives reliable information about their resistance status with a high accuracy of 88% (Table 2). That this decision protocol misses some resistant populations might be caused by temporary decrease of F_v/F_m in populations which later recovered from herbicide treatment. It is obvious that this in-season in-field herbicide resistance classification is only applicable if untreated control plants are present, which is typically not the case under practical field conditions. Moreover, herbicide resistance is estimated on a population level without precise information about the frequency of resistant plants. To overcome these limitations, the F_v/F_m values of individual sensitive and resistant plants were examined.

Herbicide sensitivity estimation on plant level

This developmental step focused on F_v/F_m values of individual plants with the objective to assess the degree of herbicide action without controls and, thus, to increase the scope of application of the PAM-imaging sensor (Wang *et al.*, 2018). At three field sites, *A. myosuroides* plants were treated with ALS, ACCase and PS II inhibitors (EXP4, Table 1). The F_v/F_m value of each single plant was assigned *post-hoc* to the two groups "resistant" and "sensitive". Per treatment, differences between these two groups were identified per measurement day using Tukey's HSD test (P < 0.05). The correct classification in resistant and sensitive plants was already possible at 3 DAT for all treatments without comparing the F_v/F_m of treated plants with untreated controls. Similar results were found by Wang *et al.* (2017) in *Apera spica-venti* (L.) Beauv. Although the assignment of plants and their respective F_v/F_m values was performed after the results on resistance were present, these experiments demonstrated the possibilities of using the PAM-imaging system on a single plant level.

Tab. 3 Utilized herbicides in experiments on herbicide induced stress in *Stellaria media* (L.) Vill. (EXP 1.1, EXP 1.2, Table 3, see text section *Classification process*). HRAC = herbicide resistance action committee, SC = suspension concentrate, SG = water-soluble granules

Trade name	Active ingredients rate	Formulation	Application rate	HRAC-group	Provider
Allie® SX®	200 g kg^{-1} metsulfuron	SG	40 g ha^{-1}	B	Du Pont de Nemours
Pointer® SX®	500 g kg^{-1} tribenuron	SG	40 g ha^{-1}	B	Du Pont de Nemours
Primus®	50 g L^{-1} florasulam	SC	0.1 L ha^{-1}	B	Dow Agroscience

---

Classification process

In the last development step, the concept of assigning (classifying) a single plant to the groups (classes) "resistant" and "sensitive" based on its F_v/F_m value was investigated. For this, a greenhouse experiment was set up as a randomized complete design and repeated twice (EXP5.1, EXP5.2, Table 1, Table 3, unpublished). Sensitive (STEME-S) and resistant (STEME-R) *Stellaria media* (L.) Vill. plants were treated in the 6-8 leave stage with three different ALS inhibiting herbicides at recommended application rates (Table 3). The F_v/F_m value of 10 plants per biotype and treatment was examined 4 DAT.

The classification into the two classes "resistant" and "sensitive" was done using a discriminant maximum likelihood classifier (Huberty & Olejnik, 2006). After training the classifier for those two classes, each measured F_v/F_m value is assigned to the most probable class. Furthermore, the training was performed with data of EXP 5.1 and the testing was done with data of EXP 5.2, and vice-versa. For each classifier, the accuracy was calculated. To compare the results of the classifiers with results of the previous data evaluation process (see text section *Herbicide sensitivity estimation on population level*), differences in the F_v/F_m value between the populations were compared using the Tukey's HSD test ($P< 0.05$).

The F_v/F_m value was always calculated as the arithmetic mean of pixels within an "area of interest". The standard deviation of pixels within an area of interest is small which makes a hypothetical effect of the standard deviation on the accuracy of the classifier unlikely. Moreover, we have found that determining the area of interest manually or automatically with background fluorescence suppressed to different degrees had only marginal effects on the F_v/F_m which supports the view that extraction of F_v/F_m from raw images yields robust data and, thus, is not a critical factor for this classification approach

Tab. 4 Maximum quantum yield of photosystem II (F_v/F_m) of sensitive (S) and resistant (R) *Stellaria media* (L.) Vill. plants 4 days after treatment in comparison with the accuracy of the discriminant maximum likelihood classifiers. The classifiers base on the single F_v/F_m values after training with data of experiment EXP 1.1 and testing with data of EXP 1.2, and vice versa (see text section *Classification process*). Values with an asterisk indicate statistically significant differences per experiment and population between untreated control and treated plants (Tukey's HSD test, P < 0.05)

Treatment	F_v/F_m				Accuracy %	
	EXP 5.1		EXP 5.2		Train EXP 5.1	Train EXP 5.2
	R	S	R	S	Test EXP 5.2	Test EXP 5.1
control	0.64	0.63	0.62	0.68	-	-
florasulam	0.62	0.53*	0.62	0.56	80	85
metsulfuron	0.65	0.58*	0.61	0.56	70	70
tribenuron	0.65	0.55*	0.64	0.51*	85	95

None of the herbicide treated STEME-S plants survived the treatment, while all STEME-R plants survived the herbicide treatments until the final visual assessment at 30 DAT. In both experiments, Tukey's HSD test (P < 0.05) attested lower F_v/F_m values of STEME-S than to STEME- R after treatment with tribenuron (Table 4). After treatment of florasulam and metsulfuron, the results were inconclusive: Tukey's HSD test discriminated STEME-S and STEME-R by their F_v/F_m value in EXP 1.1 but not in EXP 1.2. The classifiers, however, assigned up to 95 % of the plants correctly (Table 4). Likewise, in the cases of florasulam and metsulfuron, the classifiers had quite high accuracies of 70 % to 85 %. Hence, the classifier appears to be superior over the decision protocol using Tukey's HSD test, particularly when means of different populations are difficult to distinguish.

Because single plant analysis is performed with the classifier, the frequency of resistant plants within the population can be determined. Thus, analyses of resistance based on the classifier contains information on the extent of resistant plants of a population which goes beyond the simple classification of entire weed populations as "sensitive" or "resistant". Of course, to become a reliable in-field tool, the classifier should work with even higher accuracies of identification under authentic outdoor conditions.

Herbicide induced stress in crops

To investigate the phytotoxicity on crop cultivars of commonly used herbicides for weed control in sugar beet (*Beta vulgaris* spp. *vulgaris*) and soybean (*Glycine max* (L.) Merr.), Weber *et al.* (2017) examined nine sugar beet and seven soybean cultivars with the PAM-imaging sensor directly after the herbicide application in the greenhouse (EXP6, Table 1). Depending on the herbicide and cultivar, the F_v/F_m values of the herbicide-treated sugar beets were reduced significantly by 15 % to 66 % compared to the untreated control at 1 DAT. In herbicide-treated soybean, the F_v/F_m values were decreased by 9 % to 74 % within 9 DAT, depending on the herbicide and cultivar. In both cases, the PAM-imaging sensor displayed the phytotoxicity more clearly than a visual phytotoxicity assessment, which was only able to spot small differences in soybean. Li *et al.*, (2017) were able to assess the phytotoxicity of soybean herbicides on soybean with the PAM-imaging sensor in the field and found differences in the phytotoxicity of different herbicides (EXP7, Table 1).

The PAM-imaging sensor was also employed to measure herbicide-induced stress in maize (*Zea mays* L.) in a greenhouse and a field experiment conducted in 2017 (EXP8, EXP9, Table 1, (unpublished results). The experiments were aimed to compare herbicide action on various cultivars and to test how various herbicides affect the same cultivar.

The greenhouse experiment was set up as a three-factorial experiment (EXP8, Table 1). The cultivars NK RAVELLO (Syngenta Seeds GmbH, Bad Salzuflen, Germany), RIDLEY (Limagrain GmbH, Edemissen, Germany), and SUSANN (Saaten-Union GmbH, Isernhagen, Germany) were examined. Pots (17.6 x 15 cm) were filled with sandy loam and one seed per pot was placed in a depth of 4.5 cm. In the four-leaf stage, the plants were sprayed in a precision application chamber with five commonly used herbicides and herbicide tank mixes for weed control in maize (Table 4). The F_v/F_m value of single plants was measured 1 – 7, and 9 DAT. Aboveground dry biomass was evaluated 29 DAT.

Tab. 5 Treatments in the experiments on herbicide induced stress in maize (*Zea mays* L.). HRAC = herbicide resistance action committee

Treatment	Active ingredients rate	HRAC-group	Application rate (g active ingredient ha^{-1})
untreated control	-	-	-
MaisTer Power	foramsulfuron 31,5 g L^{-1}	B	47.3
	thiencarbazone 10 g L^{-1}	B	15
	iodosulfuron 1 g L^{-1}	B	1.5
Laudis +	tembotrione 44 g L^{-1}	F2	88
Spectrum	dimethenamid 720 g L^{-1}	K3	900
Kelvin OD	nicosulfuron 40 g L^{-1}	B	32
Spectrum	dimethenamid 720 g L^{-1}	K3	576
Maran	mesotrione 100 g L^{-1}	F2	80
B 235	bromoxynil 235 g L^{-1}	C3	94
Elumis +	mesotrione 75 g L^{-1}	F2	93.75
	nicosulfuron 30 g L^{-1}	B	37.5
Peak	prosulfuron 750 g kg^{-1}	B	15
Arigo +	mesotrione 360 g kg^{-1}	F2	90
	nicosulfuron 120 g kg^{-1}	B	30
	rimsulfuron 30 g kg^{-1}	B	7.5
Activus SC +	pendimethalin 400 g L^{-1}	K1	1000
B 235	bromoxynil 235 g L^{-1}	C3	58.75

The field experiment was set up as a two-factorial experiment (EXP9, Table 1). The cultivar RIDLEY was seeded at 17th of May. The same herbicides as used in the greenhouse experiment were applied with a motorized field plot sprayer (Schachtner Gerätetechnik, Ludwigsburg, Germany) in the four-leaf stage of maize. The F_V/F_m values of 7 plants per plot were assessed at similar DAT as described in the greenhouse experiment. Aboveground dry biomass was assessed 25 DAT.

In both experiments, a decrease of the F_V/F_m value of treated plants was dependent on the utilized herbicides. Especially, herbicide mixtures containing active ingredients with the mode of action in the PS II decreased the F_V/F_m values the first few DAT (Fig. 3, squares). Then, the F_V/F_m values recovered with increasing DAT but did not reach the values of the untreated control. In the greenhouse experiment, no differences in the F_v/F_m values between the three cultivars could be observed. However, all herbicide treated plants had a reduced dry biomass compared to the untreated control. In the field experiment, the untreated control had the lowest biomass, while the biomass between herbicide treated maize did not differ. Possibly, the high weed pressure in the untreated control plots lead to this biomass reduction in herbicide untreated maize plants. Based on the F_v/F_m data, mixtures without active ingredients interacting directly with the PS II seem to handle maize plants more considerately, while the results on biomass could not emphasize this..

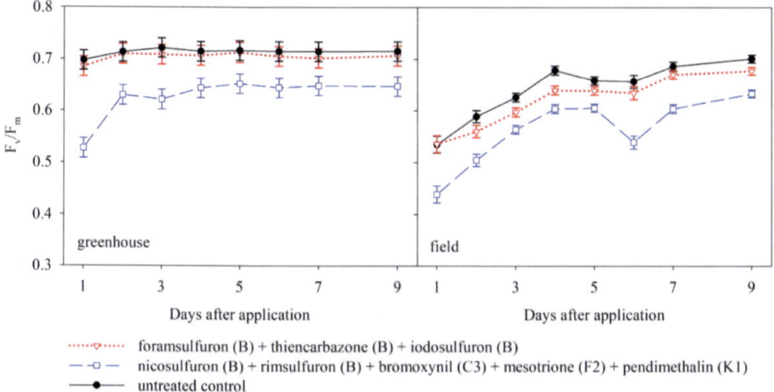

Fig. 3 Variation of maximum photochemical quantum yield of photosystem II (F_v/F_m) of maize (*Zea maize* L.) after herbicide treatment. Herbicide mode of action is displayed after each herbicide in accordance with classification by the herbicide resistance action committee

Weed competition and fungi infection

A two-factorial pot experiment with 3 repetitions was conducted in the greenhouse in 2013 to investigate if the PAM-imaging sensor was capable to detect early infestation of powdery mildew (*Blumeria graminis* f. sp. *tritici*) and weed competition in winter wheat (*Triticum aestivum* L.), cv. Toras (EXP10, Table 1, unpublished results). Winter wheat was sown at November 07 with 60 seeds in plates of 60 x 40 cm filled with a sandy loam. Plants were cultivated at temperatures of 15 °C during the day (12 h) and 12 °C at night (12 h). Powdery mildew was spray-inoculated at November 26 at 3-leaf stage of wheat with 200,000 spores/ml. Weed competition was initiated with charlock (*Sinapis arvensis* L.) sown at a density of 50 seeds per plate simultaneously with winter wheat, resulting in an average of 39 emerged weed seedlings per plate.

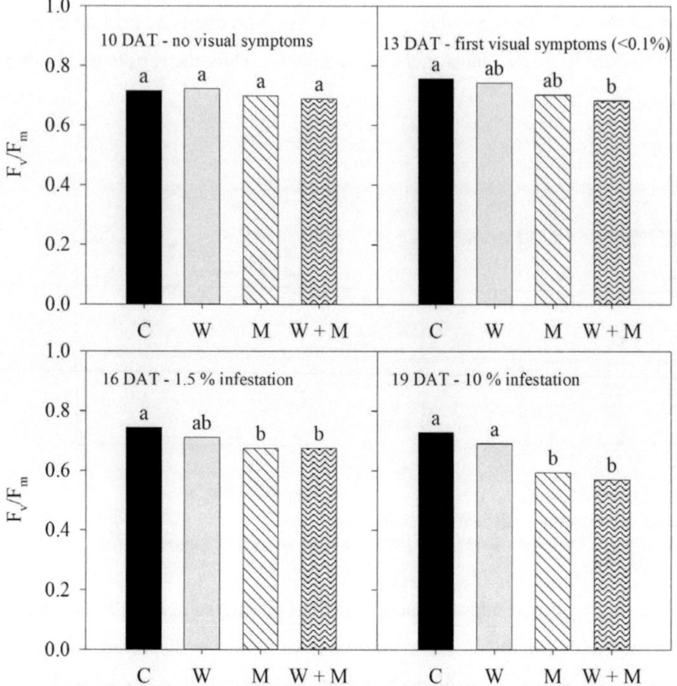

Fig. 4 Maximum photochemical quantum yield of photosystem II (F_v/F_m) of winter wheat (240 plants/m²). C = untreated control; W = winter wheat in competition with *Sinapis arvensis* L. (200 plants/m²); M = winter wheat infested with powdery mildew; M + W = winter wheat in competition with *Sinapis arvensis* L. (200 plants/m²) + winter wheat infested with powdery mildew; DAT = days after inoculation with powdery mildew

Visual assessments of disease infestation and F_v/F_m measurement (four measurements per plate) with the PAM-imaging sensor were carried out over a period of 30 days after inoculation. Plants were dark adapted for 20 minutes before measurement. The first symptoms appeared 13 DAT at 5-leaf stage with an estimated rate of less than 0.1 % infestation. Infestation rate increased to

1.5 % until 16 DAT and 10 % until 19 DAT (first tiller appeared). The combination of weed competition and powdery mildew infection resulted in significantly lower F_v/F_m values already 13 DAT compared to untreated control plants, when almost no visual symptoms appeared (Fig. 4). Powdery mildew treatments and the combination of both stresses resulted in significantly lower F_v/F_m values at 16 and 19 DAT (Fig. 4). Our data supports the view that the PAM-imaging sensor is capable for early detection of fungal disease.

2.1.5 Limitations

PAM fluorometers excite the chlorophyll fluorescence which they are measuring. Therefore, our PAM-imaging sensor is an active sensor which is only capable to assess plant stress in a narrow area of view with a close distance to the target. This excludes remote screening of stress by mounting the sensor on aerial vehicles. Screening of larger areas by a PAM sensor is possible, however, by larger series of measurements distributed over the area in accordance to established sampling patterns.

Chlorophyll fluorescence, especially the F_v/F_m value, responds to plant stress of many origins (Rosenqvist & van Kooten, 2003). Therefore, it is difficult to unequivocally trace back a decreased value for F_v/F_m to a distinct stressor. The association between stress origin and fluorescence is possible, however, if the experiment includes untreated references and obeys the *ceteris paribus* principle.

In the field, agricultural plants are exposed to changing temperatures. Pospíšil *et al.* (1998) examined chlorophyll fluorescence changes in dark acclimated leaves of winter barley (*Hordeum vulgare* L. cv. Akcent) in a laboratory experiment. They found that the F_v/F_m value was always close to 0.8 in the temperature range from -10 °C to 30 °C. Usually, field measurements are performed in this range so that pure temperature effects on F_v/F_m can be disregarded.

The situation is different when plants at various temperatures are exposed to sunlight. In the presence of strong light, low or freezing temperatures can induce photoinhibition of PS II which leads to a decreased F_v/F_m value (Krause, 1994). Hence, photoinhibition by light is a potential factor reducing the F_v/F_m in addition to herbicide stress. As the effects of photoinhibition are difficult to remove from measured F_v/F_m data, experiments testing herbicide stress should be performed under non- photoinhibitory conditions.

2.1.6 Perspectives

For future applications, the PAM imaging device is optimized for field and greenhouse studies. While this sensor is a complex tool and an expertise is needed to perform correct measurements and evaluate the data gathered, the PAM imaging sensor has to be seen as an expert tool used by scientists or high-level consultants. Nevertheless, the device has various applications covering integrated weed management, breeding, and the development of agrochemicals. In general, it can be used in the field to identify early and objective stress in the target plant. A further improvement of the classifier for herbicide resistance detection in weeds could make the device a reliable in-season tool for herbicide efficacy estimations. In breeding, the sensor could help to select the most favorable cultivar regarding stress tolerance (e. g. herbicide), while for the production of agrochemicals, the effect on plant health status of herbicides, tank mixes, adjuvants, and plant tonics could be evaluated in the field

2.1.7 Conclusion

The present study reviews the development of a PAM imaging device for stress measurements in agricultural plants. The hardware and accessories of the PAM-imaging sensor have been optimized for experiments in the greenhouse and the field. By measuring the simple parameter F_v/F_m, the PAM-imaging sensor can not only detect the action of herbicides inhibiting PS II but also of herbicides acting on distant biochemical pathways. Moreover, the F_v/F_m also responds to fungi infestation which extends the scope of application of the PAM-imaging sensor. These studies suggest that the PAM-imaging sensor can generally be utilized by breeders and growers to optimize the selectivity of herbicides. Compared to established visual methods, PAM analysis is superior because herbicide stress is analyzed quantitatively by the F_v/F_m, stress response faster, and it is not biased by the evaluating person. Future work should not only continue to explore the capacity for stress detection of PAM-imaging sensor but also soundly evaluate its economic value by comparing investments and benefits for each possible application.

Acknowledgements

This project was supported by German government's Special Purpose Fund held at Landwirtschaftliche Rentenbank. We thank Jörg Kolbowski for the software improvements. Furthermore, we thank Daniel Seitz, Melina Becker and Johannes Schlichting for their great work. A very special thank is due to Alexandra Heyn for her excellent support.

2.2 Detecting Herbicide-resistant *Apera spica-venti* with a Chlorophyll Fluorescence Agar Test

Alexander I. Linn[1,*], Pavlína Košnarová[2], Josef Soukup[2], Roland Gerhards[1]

[1]Department of Weed Science, Institute of Phytomedicine, University of Hohenheim, Stuttgart, Germany

[2]Department of Agroecology and Biometeorology, Faculty of Agrobiology, Food and Natural Resources, Czech University of Life Sciences Prague, Prague, Czech Republic

Published 2018 in: Plant, Soil and Environment, 64, 386-392; ISSN 1214-1178; Available at: https://doi.org/10.17221/110/2018-PSE

2.2.1 Abstract

Reliable tests for herbicide resistance detection are important for resistance management. Despite well-established greenhouse bioassays, faster and in-season screening methods would aid in more efficient resistance detection. The feasibility of a chlorophyll fluorescence agar-based test on herbicide resistance in *Apera spica-venti* L. was investigated. Herbicide-resistant and sensitive *A. spica-venti* seedlings were transplanted into agar containing pinoxaden and pyroxsulam herbicides. Chlorophyll fluorescence was measured and the maximum quantum efficiency of photosystem II (F_v/F_m) was determined 48 h and 72 h after the transplantation to agar, respectively. The F_v/F_m values decreased with increasing herbicide concentration. Dose-response curves and respective ED_{50} values (herbicide concentration leading to 50% decrease of the F_v/F_m value) were calculated. However, each experiment repetition exhibited different sensitivities of the populations for both herbicides. In certain cases, resistant populations demonstrated similar F_v/F_m values as sensitive populations. Contrary to the findings in *Alopecurus myosuroides* Huds., discrimination of sensitive and resistant *A. spica-venti* populations was not feasible. An increased importance of the assessment time due to the herbicide concentrations calibrated for fast responses was assumed in this study.

Keywords ALS inhibitor; ACCase inhibitor; herbicide-resistant weed; rapid detection; laboratory study; PAM fluorimeter

2.3 In-field Classification of Herbicide-resistant *Papaver rhoeas* and *Stellaria media* Using an Imaging Sensor of the Maximum Quantum Efficiency of Photosystem II

Alexander I. Linn[1]*, Robin Mink[1], Gerassimos Peteinatos[1], Roland Gerhards[1]

[1]Institute of Phytomedicine, Department of Weed Science, University of Hohenheim, Stuttgart, Germany.

Published 2019 in: Weed Research 59, 357-366; ISSN 1365-3180;

Available at: https://doi.org/10.1111/wre.12374

2.3.1 Summary

Reliable in-season and in-field tools for rapidly quantifying herbicide efficacy in dicotyledonous weeds are missing. In this study, the maximum quantum efficiency of photosystem II (F_v/F_m) of susceptible and resistant *Papaver rhoeas* and *Stellaria media* populations in response to treatments with acetolactate synthase (ALS) inhibitors were examined. Seedlings (4–6 leafs) were transplanted into the field immediately after the application of the ALS inhibitors florasulam, metsulfuron-methyl and tribenuron-methyl. The F_v/F_m values were assessed 1–7, 9 and 14 days after treatment (DAT). Based on the F_v/F_m values of all fluorescing pixels in the images of herbicide-treated plants, discriminant maximum-likelihood classifiers were created. Based on this classifier, an independent set of images were classified into 'susceptible' or 'resistant' plants. The classifiers' accuracy, false-positive rate and false-negative rate were calculated. The F_v/F_m values of sensitive *P. rhoeas* and *S. media* plants decreased within 3 DAT by 28–43%. The F_v/F_m values of the resistant plants of both species were 20% higher than those of the sensitive plants in all herbicide treatments. The classifier separated sensitive and resistant plants 3 DAT with accuracies of 62–100%. With increasing DAT, the false-negative and false-positive classifications decreased. We conclude that by the assessment of the F_v/F_m value in combination with the classification sensitive and resistant *P. rhoeas* and *S. media* populations could be separated 3 DAT. This technique can help to select effective control methods and speed up the monitoring process of susceptible and resistant weeds.

Keywords: ALS inhibitor; decision support system; discriminant analysis; pulse amplitude modulated imaging fluorometer; herbicide resistance management

Chapter III

General Discussion

3 General Discussion

In this study, the development of a mobile pulse amplitude modulated (PAM)-imaging chlorophyll fluorometer for field applications has been explored characterizing the physical properties, scope of application, and limitations of the device. Further, the maximum quantum efficiency of PS II (F_v/F_m) measured by PAM-imaging devices was investigated to derive information on herbicide resistance in *Apera spica-venti* L. Beauv, *Stellaria media* L. Vill, and *Papaver rhoeas* L. Laboratory, greenhouse, and field studies have been performed. The respective results are illustrated in three papers as a part of this thesis. This chapter summarizes and discusses the outcomes of the investigations and appends future perspectives.

3.1 Chlorophyll Fluorescence in Agriculture

Measuring chlorophyll fluorescence is widespread in physiological and ecophysiological studies to examine photosynthetic performance and stress in plants (Rosenqvist & van Kooten, 2003; Baker, 2008). In the topic of weed science, chlorophyll fluorescence studies have been performed to quantify herbicide action on weeds and to derive information on the mode of herbicide action (Barbagallo *et al.*, 2003; Riethmuller-Haage *et al.*, 2006). With the rise of herbicide-resistant weeds, the need for tools and screening protocols to detect herbicide resistance increased. This lead to the development of various screening protocols and tests on herbicide resistance (Beckie *et al.*, 2000; Burgos *et al.*, 2013). Until today, the most common screening method to identify herbicide-resistant weed populations are pot bioassays (Burgos, 2015). Viable seeds of the population to be investigated are collected in the field. This is usually done in-season when a lack of efficacy becomes apparent after an unsatisfactory herbicide application and before crop harvest. After collecting the seeds, the plants are grown from the seeds in soil. Commonly one discriminating dose of the herbicides to be tested is applied. The evaluation of the herbicide efficacy depends on the herbicide employed and is performed approximately 28 days after application. This is done by comparing visual damage symptoms, growth reduction or/and fresh or dry biomass with a sensitive reference and untreated control plants. Additionally, the survival rate is recorded. This screening method is a reliable procedure (Streibig, 1988). The results can be transferred to the R-rating scale, which assigns populations by the assessed herbicide efficacy to different resistance classes (Moss *et al.*, 2007). This scaling gives a simplified overview of the resistance status of the population and easies a comparison of different populations. Besides the valuable information which greenhouse pot bioassays can provide, a major disadvantage of these bioassays is the rather long duration until results become available. The trial period from seeding until the completion of the pot bioassay easily exceeds 2 months. Therefore, targeted actions to curb the number of resistant plants can take action only in the next season. Hence, resistance management at an early stage is not possible. Due to the late availability of the screening results, there is no time left for an adjustment of the management practices for the current season, if necessary. Therefore, during the cultivation period, where weed control measures may have already been performed, it is not feasible to apply correction measures. Furthermore, a delay of the herbicide application until the screening results are accessible is not a sufficient solution, as the weed control efficacy of herbicides decreases with increasing weed growth stages. Fast and reliable identification of resistant weeds is imperative. Since the duration of screenings on resistance is one of the most important factors that needs to be improved, solutions for timely identification of herbicide-resistant weeds need to be developed. Laboratory and greenhouse screenings need to be speeded up while ensuring the accuracy of the results on herbicide resistance. In this thesis, several attempts have been performed using PAM-imaging chlorophyll fluorometers to enable a fast and secure estimation on herbicide sensitivity.

The assessment of chlorophyll fluorescence and chlorophyll fluorescence parameters provides information on the photosynthetic performance. In herbicide treated plants, changes in the performance of the photosynthetic apparatus may appear significantly much faster, than damage symptoms become visible. The first studies employing chlorophyll fluorescence parameters for herbicide resistance detection were performed with respect to photosystem II (PS II) inhibitors (Ahrens *et al.*, 1981; Ali & Machado, 1981; Vencill & Foy, 1988; van Oorschot & van Leeuwen, 1992). Here, sensitive dicotyledonous and monocotyledonous weeds were identified by increased relative fluorescence intensity in light conditions within a few hours after floating detached leaf parts with a solution containing discriminating herbicide dosage. Voss *et al.* (1984) found out that a dark acclimation is necessary for the reproducibility of the results. These early studies on herbicide resistance detection were performed with non-PAM-imaging chlorophyll fluorometers, which had a rather complicated setup. Usually, the sensors were permanently installed and only available in special laboratories. Though the fluorescence ratio F_v/F_m has been identified to represent the maximum quantum efficiency of PS II already in 1975 (Kitajima & Butler, 1975), at this time the technical restrictions did not allow an intensive usage of this fluorescence ratio. The commercial introduction of PAM fluorometers made it possible for numerous laboratories to access this technique and thus enabled intensive and comprehensive physiological and ecophysiological plant studies (Schreiber *et al.*, 1995). In a laboratory study, the detection of acetolactate synthase (ALS) and acetyl-Coa carboxylase (ACCase) resistant *Alopecurus myosuroides* Huds. populations has been realized with a PAM-imaging fluorometer by employing the chlorophyll fluorescence ratio F_v/F_m in dark acclimated plants (Kaiser *et al.*, 2013). This showed the possibility to differentiate between resistant and sensitive *A. myosuroides* plants with a PAM-fluorometer within a few hours. This PAM-imaging chlorophyll fluorometer was laboratory equipment, which did not have any portable capabilities. Nevertheless, the advantage of a fast discrimination between resistant and sensitive *A. myosuroides* populations triggered investigations of this technique in field studies. For applications of the PAM-imaging technique in the field, a portable device needed to be created without losing the precision of the permanently installed device. In the laboratory, plants can be easily dark-acclimated prior to the measurements by switching off the light. On the other hand, this is not practicable for field measurements, therefore dark-acclimation of the target plants needs to be achieved with a different setup in the field. This is an example of technical restrictions and practical problems that need to be overcome in the field. A mobile PAM-imaging chlorophyll fluorometer prototype has been designed and satisfactorily tested for an in-season detection of herbicide-resistant *A. myosuroides* in the field (Wang *et al.*, 2016). Since that, the PAM-imaging device underwent steadily further development and adjustment meanwhile further use cases of the system for herbicide sensitivity estimations have been identified.

The review paper "Development and applications of a field imaging chlorophyll fluorometer to measure stress in agricultural plants" gives a detailed description of the portable PAM-imaging fluorometer, suggests possible applications based on published research and new case studies, and describes system-related limitations. The device has been used for the determination of the chlorophyll fluorescence ratio F_v/F_m, which is a precise indicator for plant stress. With respect to herbicide resistance detection, the major study object was *A. myosuroides*. Here, herbicide-resistant populations and single resistant plants can be identified in-season with the PAM-imaging sensor. Apart from resistance detection, the PAM-imaging system can also be applied in other areas of agriculture. In another research question, the PAM-imaging device has been proved to estimate herbicide damage in crops, which enables the optimization of herbicide mixtures and cultivars. The system described can be employed in crops and weeds to detect plant stress induced by herbicides not only interacting directly with the PS II, but also by herbicides with mode of action in biochemical paths distant from PS II. Hence, the origin of the stress cannot be pinpointed by F_v/F_m data alone. However, the stress reported by F_v/F_m can be clearly attributed to the herbicide treatment, if compared to an untreated control. The scope of application in crop monitoring and herbicide resistance detection of the portable PAM-imaging device is not fully exploited, yet. There is an urgent need to expand the use of the PAM-imaging device for herbicide resistance detection in more weed species in order to utilize it in its full potential. In this thesis, two further scopes of application are elucidated to investigate the PAM-imaging fluorometer for herbicide resistance detection. On the one hand, a laboratory study was performed to adjust a fast screening protocol on herbicide resistance in *A. spica-venti*. On the other hand, the portable PAM-imaging fluorometer has been investigated for the detection of herbicide resistance in *S. media* and *P. rhoeas* to ALS inhibitors in the field, including an automated classification of single plants to the classes "resistant" and "sensitive".

3.2 Detection of Herbicide-resistant *Apera spica-venit*

Apera spica-venti is an important winter annual monocotyledonous weed in Northern, Eastern and Central Europe (Hamouzová *et al.*, 2011). It mainly infests winter cereal fields such as winter wheat and winter barley, but can also cause yield losses in oilseed rape, forage crops and early-sown spring cereals (Gerowitt & Heitefuss, 1990). Short crop rotations, an increasing percentage of winter cereals in the cropping system, and reduced-/no-tillage practices have resulted in an increased presence of *A. spica-venti* (Massa & Gerhards, 2011). *Apera spica-venti* competes especially for light with the crop by overgrowing it, causing up to 30% yield losses at a density of 200 plants per m^2 (Melander, 1995; Melander *et al.*, 2008). Herbicides are used most commonly for the targeted control of *A. spica-venti*. However, the repeated use of herbicides with the same mode of action has promoted the evolution of resistant populations (Soukup *et al.*, 2006). Until today, *A. spica-venti* populations with resistance to ALS, ACCase and PS II inhibiting herbicides have been reported in 10 European countries (Heap, 2019). Though the total number of herbicide-resistant *A. spica-venti* populations cannot be stated, herbicide-resistant *A. spica-venti* populations are already widely distributed, and a more extensive occurrence is expected. To identify herbicide-resistant *A. spica-venti* populations, it is of high importance to find and establish fast and robust screening methods for herbicide resistance confirmation. The target of this investigation was to establish a fast and reliable laboratory screening for the detection of herbicide resistance in *A. spica-venti* by using PAM-imaging chlorophyll fluorometry. Besides a fast availability of results, the new screening method had to achieve reproducible and comparable results.

The second paper of this thesis entitled "Detecting herbicide-resistant *Apera spica-venti* with a chlorophyll fluorescence agar test" investigates the feasibility of an agar test on herbicide resistance in *A. spica-venti* by utilizing PAM-imaging chlorophyll fluorometry. Herbicide-resistant and sensitive *A. spica-venti* seedlings were transplanted into agar treated with pinoxaden and pyroxsulam. The F_v/F_m value was determined 48 h and 72 h after transplanting to pinoxaden and pyroxsulam treated agar, respectively. Based on the F_v/F_m values dose-response curves were fitted and ED_{50} (dosage leading to 50% decrease in F_v/F_m) values calculated. The F_v/F_m values decreased with increasing herbicide concentration. The F_v/F_m values clearly indicated the plant stress induced by the herbicides, depending on the herbicide concentration. Nevertheless, the results of the agar-based test system could not clearly state the herbicide resistance classification derived from the greenhouse assays due to high variabilities in the calculated ED_{50} values. The F_v/F_m based approach, therefore, did not lead to fully satisfactory results. In the described method, herbicide caused plant stress was detectable very fast, which was the main aim. Though this emphasizes the fastness of herbicide sensitivity estimation by the determination of the F_v/F_m value, the variability in the evaluation on resistance between the experiments does not represent a reliable outcome. It has to be stated that the current protocol cannot be utilized for the secure identification of herbicide-resistant *A. spica-venti* and the procedure needs to be reconsidered.

The performance of a herbicide strongly depends on the environmental conditions including factors such as temperature, light, and soil (Kudsk & Streibig, 2003). To minimize the variation of light, temperature, and soil, in the presented approach the experiments were performed in a climate chamber with fixed settings, and soil was replaced by agar. Beside the environmental conditions, the plant response depends on the applied herbicide concentration, duration of exposition, time point of measurement, and the plant itself (Streibig, 2003). In the presented approach, the herbicide concentrations used ranged up to a multiple of the recommended field rates in order to provoke a fast plant reaction. In addition, plants were exposed to the herbicides steadily from transplanting until chlorophyll fluorescence measurement. As the herbicide concentration increases, the time required to observe the same response decreases. This is also indicated by the calculated dose-response curves. By implication, the plant response changes faster when the herbicide concentration increases. Thus, it might be the case, that the herbicide concentrations used here, accelerated the plant response so strongly that a point of measurement at hours scale was not sufficient enough to elaborate differences in the plant response of resistant and sensitive plants. In consequence, a more precise resolution of the time point of measurement could serve results that are more robust. Nevertheless, the practicability of a better time resolution has to be proven and might be hard to be achieved. As performed in this study, dose-response experiments are a common setup to estimate the strength of resistance (Streibig, 1988; Nielsen *et al.*, 2004; Knezevic *et al.*, 2007). Time can be incorporated into such experiments as an additional factor. This can be of particular advantage when investigating new screening approaches for resistance detection. In order to discriminate between resistant and sensitive populations, it is essential to know the herbicide dosage that identifies sensitivity at a given time. The creation of time-dose-response relations gives a more comprehensive overview of differences between resistant and sensitive populations regarding herbicide concentration and time relations than sole dose-response experiments. To elaborate a more robust screening protocol, the interplay of time, herbicide concentration, and the F_v/F_m value in combination with the used sensitive and resistant references needs a deeper investigation. In case, a more precise resolution of the time of measurement cannot be achieved, the herbicide concentrations need to be decreased. Then, the impact of the factor time decreases, as plant health status will change slower. This automatically leads to a longer trial period. That contradicts with the target of a fast screening method. A tradeoff between time and accuracy needs to be further researched and examined. The further advancement of the described approach would benefit by including time-dose-response experiments. Then the herbicide dosages and time points of measurement indicated by significant differences between F_v/F_m values of sensitive and resistant populations could be determined. The fast and non-destructive determination of the F_v/F_m value, provided by the PAM-imaging technology, is well suited for this kind of study.

3.3 In-field Identification of Herbicide-resistant *Papaver rhoeas* and *Stellaria media* Plants

The monitoring and evaluation of the herbicide efficacy is an important component of herbicide resistance management in plant protection (Beckie & Harker, 2017). Usually, the herbicide efficacy is assessed when visual symptoms of the herbicide damage have appeared at the target plants. However, the timing of these symptoms depends on several factors including the herbicide application efficacy, weed development stage and age, weather conditions, and herbicide mode of action. An early identification of an unsatisfactory herbicide efficacy may allow sufficient time for additional weed control measures. After a herbicide application, plant stress can be detected by the determination of chlorophyll fluorescence ratio F_v/F_m before herbicide symptoms occur, and thereby accelerating the monitoring process. The F_v/F_m value has been investigated for herbicide resistance detection in monocotyledonous weeds with resistance to ACCase-, ALS-, and PS II-inhibiting herbicides (Wang *et al.*, 2016; Zhang *et al.*, 2016; Wang *et al.*, 2018). In dicotyledonous weeds, only non-PAM-imaging chlorophyll fluorometers have been used for herbicide resistance studies with a focus on PS II inhibiting herbicides (Ahrens *et al.*, 1981; Ali & Machado, 1981). The suitability of the F_v/F_m value for the detection of ALS herbicide-resistant dicotyledonous weeds has not been investigated, yet. Recent studies with *A. myosuroides*, *Beta vulgaris* spp. *vulgaris* and *Glycine max* (L.) Merr. have shown, that the time of F_v/F_m measurement for herbicide sensitivity estimations differ by species and herbicide (Roeb *et al.*, 2015; Wang *et al.*, 2016; Weber *et al.*, 2017; Wang *et al.*, 2018). This indicates that general conclusions on the time for herbicide sensitivity estimations by F_v/F_m cannot be drawn and have to be determined per species and herbicide specific. Therefore, it is uncertain whether herbicide sensitivity estimations in *S. media* and *P. rhoeas* can be performed by F_v/F_m measurements, nor when the best time for this valuation is. The present study, therefore, investigated the possibility of performing F_v/F_m measurements in *S. media* and *P. rhoeas* for herbicide sensitivity estimations.

The third paper of this thesis entitled "In-field classification of herbicide-resistant *Papaver rhoeas* and *Stellaria media* using an imaging sensor of the maximum quantum efficiency of photosystem II" describes the employment of the F_v/F_m value for the detection of herbicide-resistant *P. rhoeas* and *S. media* populations to ALS inhibitors using a portable PAM-imaging chlorophyll fluorometer. Herbicide sensitive and resistant *P. rhoeas* and *S. media* plants were transplanted into the field immediately after treatment with metsulfuron-methyl, tribenuron-methyl, and florasulam. The F_v/F_m value of each individual plant has been determined at days from 1 to 7, 9 and 14 after treatment. Within 3 days after treatment (DAT), the F_v/F_m values of herbicide treated sensitive *P. rhoeas* and *S. media* plants decreased by 28% to 48%, respectively, compared to the untreated control for all herbicide treatments. At the same time after metsulfuron-methyl and tribenuron-methyl treatment, the F_v/F_m values of resistant *P. rhoeas* and *S. media* plants showed no decrease compared to the untreated control. However, after florasulam treatment, resistant plants of both species had a decrease in their F_v/F_m value. The F_v/F_m values indicated strong plant stress in sensitive plants and no (metsulfuron-methyl, tribenuron-methyl) or less (florasulam) stress in resistant plants. This allowed a clear discrimination of the mean F_v/F_m values of resistant and sensitive populations with Tukey's HSD test ($P < 0.05$) after metsulfuron-methyl and tribenuron-methyl treatment. After the florasulam treatment, the discrimination was not clear at several days and the resistant populations may have been identified as apparent sensitive due to the present herbicide stress.

The results for the florasulam treatment indicate the intrinsic problems incorporated when herbicide resistance is attested with a decision protocol that certifies sensitivity by statistical significance compared to the untreated control. Due to a significant stress in the resistant *S. media* and *P. rhoeas* plants after florasulam treatment, sensitivity would be attested for both the sensitive and the resistant populations, which is explicitly wrong. Therefore, at best, the comparison should take place only between the F_v/F_m values of herbicide treated plants in combination with a predefined database independent from the untreated control. For this, based on the measured F_v/F_m values of herbicide treated plants, maximum likelihood classifiers, assigning single plants to the classes "susceptible" or "resistant", were created and tested with independent data sets. The classifier identified single susceptible and resistant *S. media* plants based on their F_v/F_m value at 3 DAT with accuracies of 88% to 100% for all three herbicides. At the same time, for *P. rhoeas*, the accuracies of the classifier were lower with 62% to 95%. These high accuracies indicate that a classification of single plants by their F_v/F_m value after herbicide treatment is possible. The classifier even can outperform a common statistical comparison. Especially for the florasulam treatment, the created classifiers were able to elaborate the differences between resistant and sensitive plants better than the statistical comparison. Furthermore, the statistical comparison can only provide the information "significant" (sensitive) or "not significant" (resistant). It is not able to provide information on the frequency of resistant individuals. The classifier, in contrast, identifies the number of resistant and susceptible specimen and, therefore, provides information on the frequency of resistant individuals within the examined population. In addition, the classifier is superior to statistical comparison of treated with untreated plants for practical reasons. In practical farming, usually, the entire field is treated with herbicides and unsprayed zones are seldom present. To evaluate the sensitivity of the target weed population by the protocol for statistical comparison, untreated field areas are necessary. Untreated zones could be created in the field, but this includes two problems. On the one hand, in the untreated area, where the herbicide application is prohibited, there will be no weed control. Therefore, in this zone a timely separated weed control measure becomes necessary, which at first can be performed after the measurement with the PAM-imaging sensor, increasing though the cost of weed management. On the other hand, an additional effort prior to the first herbicide application is needed to identify the proper area serving as an untreated zone as the species of interest definitively needs to be present in sufficient amount. In addition, to reduce variability in the results due to other factors than the herbicide treatment but maybe affecting the herbicide efficacy, heterogeneity within the samples for measurement has to be maximised. Especially, when scouting large fields, various untreated areas may become necessary in order to take the existing spatial variability into account. And in each untreated area, a sufficient number of untreated plants has to be measured. This results in a considerable input which has to be delivered in addition to the measurements themselves. Contrary, the screening protocol with the classifier does not need an additional effort prior to the actual measurements. Due to the setup of the classifier, no untreated control is required since the assignment of the single plants does not include F_v/F_m values of untreated plants as reference. This makes the arrangement of untreated zones redundant. It is obvious that the classifier can just give a probability of resistance, based on the previous knowledge, and the current measurements. But based on that information, a more accurate decision about the resistance level in the field can be drawn and decided if correction treatments are necessary.

This field study was performed with characterized populations, which were pre-grown in the greenhouse, treated with herbicide in the application chamber, and afterwards transplanted into the field. Therefore, the study has a few characteristics of an idealized case study. To validate the PAM-imaging sensor and the classifiers for the detection of herbicide-resistant *S. media* and *P. rhoeas* plants in practical conditions, further experiments are necessary. At several fields, field-own populations have to be measured with the PAM-imaging sensor after herbicide treatment and examined if the assignments of the classifiers are correct. Meanwhile, the accuracies have to be increased as the current identification rate is too low for secure estimations, especially in *P. rhoeas*. Furthermore, the resistant populations investigated had a highly to moderately qualified resistance. It can be expected that the classification based on the F_v/F_m value has its limit at low-quality resistance. This is when plants are heavily damaged by the herbicide and seem to die but recover with time. When measuring the plant stress 3 DAT, these resistant plants with low resistance might show the same F_v/F_m values as sensitive plants. It needs to be investigated up to which strength of resistance the classifier is still capable to produce secure results. In addition, it is necessary to test different herbicide concentrations in order to know the plant response at low concentrations. Here, sensitive plants are of particular interest to determine the herbicide concentrations at which sensitive plants show a similarly low change in the F_v/F_m value as resistant plants. This could further estimate the limits of the PAM-imaging sensor for herbicide resistance detection and, therefore, ensure the safety of results.

Overall, the PAM-imaging sensor enables a fast and non-destructive plant status determination in agricultural plants. The employment of the F_v/F_m value has been proven a good estimator for plant stress due to herbicides in crops and weeds. Screening on herbicide resistance by F_v/F_m in *A. spica-venti* requires a more comprehensive study to elaborate the ideal herbicide concentrations and points of time to detect differences in resistant and sensitive populations. In the field, *P. rhoeas* and *S. media* populations with strong herbicide resistance can be identified by comparing the mean F_v/F_m value of the examined plants after herbicide treatment with an untreated control. Further, an automated classification of single *P. rhoeas* and *S. media* plants to the classes "resistant" or "susceptible" based on their measured F_v/F_m values after herbicide treatment is possible and may outperform a common statistical comparison. After increasing the reliability of the results, the PAM-imaging sensor can contribute significantly to the monitoring of herbicide efficacy and herbicide resistance detection.

4 General References

ABBASPOOR M & STREIBIG JC (2005) Clodinafop changes the chlorophyll fluorescence induction curve. *Weed science* **53,** 1–9.

AHRENS WH, ARNTZEN CJ & STOLLER EW (1981) Chlorophyll fluorescence assay for the determination of triazine resistance. *Weed science* **29,** 316–322.

ALI A & MACHADO VS (1981) Rapid detection of 'triazine resistant' weeds using chlorophyll fluorescence. *Weed Research* **21,** 191–197.

BAKER NR (2008) Chlorophyll fluorescence: A probe of photosynthesis in vivo. *Annual review of plant biology* **59,** 89–113.

BARBAGALLO RP, OXBOROUGH K, PALLETT KE & BAKER NR (2003) Rapid, noninvasive screening for perturbations of metabolism and plant growth using chlorophyll fluorescence imaging. *Plant Physiology* **132,** 485–493.

BAURIEGEL E & HERPPICH W (2014) Hyperspectral and chlorophyll fluorescence imaging for early detection of plant diseases, with special reference to *Fusarium* spec. infections on wheat. *Agriculture* **4,** 32–57.

BECKIE HJ & HARKER KN (2017) Our top 10 herbicide-resistant weed management practices. *Pest management science* **73,** 1045–1052.

BECKIE HJ, HEAP IM, SMEDA RJ & HALL LM (2000) Screening for herbicide resistance in weeds. *Weed technology* **14,** 428–445.

BELFRY KD, SOLTANI N, BROWN LR & SIKKEMA PH (2015) Tolerance of identity preserved soybean cultivars to preemergence herbicides. *Canadian journal of plant science* **95,** 719–726.

BIGOT A, FONTAINE F, CLÉMENT C & VAILLANT-GAVEAU N (2007) Effect of the herbicide flumioxazin on photosynthetic performance of grapevine (*Vitis vinifera* L.). *Chemosphere* **67,** 1243–1251.

BLACKBURN GA (2006) Hyperspectral remote sensing of plant pigments. *Journal of Experimental Botany,* **58,** 855-867.

BURGOS NR (2015) Whole-plant and seed bioassays for resistance confirmation. *Weed science* **63,** 152–165.

BURGOS NR, TRANEL PJ, STREIBIG JC et al. (2013) Confirmation of resistance to herbicides and evaluation of resistance levels. *Weed science* **61,** 4–20.

FOLLAK S & HURLE K (2004) Recovery of non-target plants affected by airborne bromoxynil-octanoate and metribuzin. *Weed Research* **44,** 142–147.

GAMON JA, PENUELAS J & FIELD CB (1992) A narrow-waveband spectral index that tracks diurnal changes in photosynthetic efficiency. *Remote Sensing of environment* **41,** 35–44.

GERHARDS R, GUTJAHR C, WEIS M, KELLER M, SÖKEFELD M, MÖHRING J & PIEPHO HP (2011) Using precision farming technology to quantify yield effects attributed to weed competition and herbicide application. *Weed Research,* **52,** 6-15.

GEROWITT B & HEITEFUSS R (1990) Requirements and possibilities for weed control according to economic thresholds in cereals. In: *Crop Protection in Northern Britain,* 55–62. Association for Crop Protection in Northern Britain.

GOVENDER M, GOVENDER PJ, IM WEIERSBYE, WITKOWSKI ETF & AHMED F (2009) Review of commonly used remote sensing and ground-based technologies to measure plant water stress. *Water Sa* **35.**

HAMOUZOVÁ K, SOUKUP J, JURSIK M, HAMOUZ P, VENCLOVÁ V & TŮMOVÁ P (2011) Cross-resistance to three frequently used sulfonylurea herbicides in populations of *Apera spica-venti* from the Czech Republic. *Weed Research* **51,** 113–122.

HANKAMER B, BARBER J & BOEKEMA EJ (1997) Structure and membrane organization of photosystem II in green plants. *Annual review of plant biology* **48,** 641–671.

HEAP I (2014) Global perspective of herbicide-resistant weeds. *Pest management science* **70,** 1306–1315.

HEAP I (2019) The international survey of herbicide resistant weeds. www.weedscience.org, last accessed 8 May 2019.

HILTON HW (1957) Herbicide tolerant strains of weeds. *Hawaiian Sugar Planters Association Annual Report,* 69–72.

HUBERTY CJ & OLEJNIK S (2006) Applied MANOVA and discriminant analysis. John Wiley & Sons.

JIN ZL, ZHANG F, AHMED ZI et al. (2010) Differential morphological and physiological responses of two oilseed *Brassica* species to a new herbicide ZJ0273 used in rapeseed fields. *Pesticide Biochemistry and Physiology* **98,** 1–8.

KAISER YI, MENEGAT A & GERHARDS R (2013) Chlorophyll fluorescence imaging: A new method for rapid detection of herbicide resistance in *Alopecurus myosuroides*. *Weed Research* **53,** 399–406.

KAUTSKY H & HIRSCH A (1931) Neue Versuche zur Kohlenstoffassimilation. *Naturwissenschaften* **19,** 964.

KITAJIMA M & BUTLER WL (1975) Quenching of chlorophyll fluorescence and primary photochemistry in chloroplasts by dibromothymoquinone. *Biochimica et Biophysica Acta* **376**, 105–115.

KNEZEVIC SZ, STREIBIG JC & RITZ C (2007) Utilizing R software package for dose-response studies: The concept and data analysis. *Weed technology* **21**, 840–848.

KNIPLING EB (1970) Physical and physiological basis for the reflectance of visible and near-infrared radiation from vegetation. *Remote Sensing of Environment*, **1**, 155-159.

KRAUSE HG (1994) Photoinhibition induced by low temperatures. In: *Photoinhibition of photosynthesis. From molecular mechanisms to the field* (eds. NR BAKER & JR BOWYER), 331–348. Bios Scientific Publishers.

KUDSK P & STREIBIG JC (2003) Herbicides – a two-edged sword. *Weed Research* **43**, 90–102.

LI H, WANG P, WEBER J & GERHARDS R (2017) Early identification of herbicide stress in soybean (*Glycine max* (L.) Merr.) using chlorophyll fluorescence imaging technology. *Sensors* **18**, 21.

LINDENTHAL M, STEINER U, DEHNE H-W & OERKE E-C (2005) Effect of downy mildew development on transpiration of cucumber leaves visualized by digital infrared thermography. *Phytopathology*, **95**, 233-240.

LOLL B, KERN J, SAENGER W, ZOUNI A & BIESIADKA J (2005) Towards complete cofactor arrangement in the 3.0 Å resolution structure of photosystem II. *Nature* **438**, 1040–1044.

LOWE A, HARRISON N & FRENCH AP (2017) Hyperspectral image analysis techniques for the detection and classification of the early onset of plant disease and stress. *Plant methods* **13**, 1–12.

MACINNIS-NG CMO & RALPH PJ (2003) Short-term response and recovery of *Zostera capricorni* photosynthesis after herbicide exposure. *Aquatic Botany* **76**, 1–15.

MASSA D & GERHARDS R (2011) Investigations on herbicide resistance in European silky bent grass (*Apera spica-venti*) populations. *Journal of Plant Diseases and Protection* **118**, 31–39.

MELANDER B (1995) Impact of drilling date on *Apera spica-venti* L. and *Alopecurus myosuroides* Huds. in winter cereals. *Weed Research* **35**, 157–166.

MELANDER B, HOLST N, JENSEN PK, HANSEN EM & OLESEN JE (2008) *Apera spica-venti* population dynamics and impact on crop yield as affected by tillage, crop rotation, location and herbicide programmes. *Weed Research* **48**, 48–57.

MOSS SR, PERRYMAN SAM & TATNELL LV (2007) Managing herbicide-resistant blackgrass (*Alopecurus myosuroides*): Theory and practice. *Weed technology* **21**, 300–309.

NIELSEN OK, RITZ C & STREIBIG JC (2004) Nonlinear mixed-model regression to analyze herbicide dose–response relationships. *Weed technology* **18**, 30–37.

OERKE EC & STEINER U (2010) Potential of digital thermography for disease control. In: *Precision crop protections – the challenge and use of heterogeneity* (eds. EC OERKE, R GERHARDS, G MENZ & RA SIKORA), 167–182.

PASCUA JAA, PRADO AJA, SOLIS BRB, CID-ANDRES AP & CAMBIADOR CJB (2019) Trends in fabrication, data gathering, validation, and application of molecular fluorometer and spectrofluorometer. *Spectrochimica Acta Part A: Molecular and Biomolecular Spectroscopy*, **220**; doi.org/10.1016/j.saa.2019.02.061.

PETEINATOS G, KORSAETH A, BERGE T & GERHARDS R (2016) Using optical sensors to identify water deprivation, nitrogen shortage, weed presence and fungal infection in wheat. *Agriculture* **6**, 24.

POSPÍŠIL P, SKOTNICA J & NAUŠ J (1998) Low and high temperature dependence of minimum F_0 and maximum F_M chlorophyll fluorescence in vivo. *Biochimica et Biophysica Acta* **1363**, 95–99.

POUDYAL D, ROSENQVIST E & OTTOSEN C-O (2019) Phenotyping from lab to field–tomato lines screened for heat stress using F_v/F_m maintain high fruit yield during thermal stress in the field. *Functional plant biology* **46**, 44–55.

RIETHMULLER-HAAGE I, BASTIAANS L, KROPFF MJ, HARBINSON J & KEMPENAAR C (2006) Can photosynthesis-related parameters be used to establish the activity of acetolactate synthase–inhibiting herbicides on weeds? *Weed science* **54**, 974–982.

ROEB J, PETEINATOS GG & GERHARDS R (2015) Using sensors to assess herbicide stress in sugar beet. In: *Precision agriculture'15* (ed. JV STAFFORD), 259–269. Wageningen Academic Publishers.

ROSENQVIST E & VAN KOOTEN O (2003) Chlorophyll fluorescence: A general description and nomenclature. In: *Practical applications of chlorophyll fluorescence in plant biology*, 31–77. Springer.

ROUSE JW, HAAS RH, SCHELL JA & DEERING DW (1973) Monitoring the vernal advancement and retrogradiation (green waved effect) of natural vegetation. *Technical Report NASA,* 112.

SALZMAN FP & RENNER KA (1992) Response of soybean to combinations of clomazone, metribuzin, linuron, alachlor, and atrazine. *Weed technology* **6**, 922–929.

SCHREIBER U, BILGER W & NEUBAUER C (1995) Chlorophyll fluorescence as a nonintrusive indicator for rapid assessment of in vivo photosynthesis. In: *Ecophysiology of photosynthesis* (eds. ED SCHULZE & MM CALDWELL), 49–70. Springer.

SMITH AE & WILKINSON RE (1974) Differential absorption, translocation and metabolism of metribuzin [4-amino-6-tert-butyl-3-(methylthio)-as-triazine-5(4H)one] by soybean cultivars. *Physiologia plantarum* **32,** 253–257.

SOUKUP J, NOVAKOVA K, HAMOUZ P & NAMESTEK J (2006) Ecology of silky bent grass (*Apera spica-venti* (L.) Beauv.), its importance and control in the Czech Republic. *Journal of Plant Disease and Protection, special issue 20,* 73–80.

STIRBET A & GOVINDJEE (2011) On the relation between the Kautsky effect (chlorophyll a fluorescence induction) and photosystem II. Basics and applications of the OJIP fluorescence transient. *Journal of Photochemistry and Photobiology B: Biology* **104,** 236–257.

STREIBIG JC (1988) Herbicide bioassay. *Weed Research* **28,** 479–484.

STREIBIG JC (2003) Assessment of herbicide effects. http://www.ewrs.org/et/docs/Herbicide_inderaction.pdf, last accessed 8 May 2019.

SWITZER CM (1957) The existence of 2, 4-D resistant strains of wild carrot. *Proc. Northeast Weed Contr. Conf.,* 315–318.

TRISSL H-W, GAO Y & WULF K (1993) Theoretical fluorescence induction curves derived from coupled differential equations describing the primary photochemistry of photosystem II by an exciton-radical pair equilibrium. *Biophysical journal* **64,** 974–988.

VADIVAMBAL R & JAYAS DS (2011) Applications of thermal imaging in agriculture and food industry – a review. *Food Bioprocess Technology,* **4,** 186-199.

VAN OORSCHOT JLP & VAN LEEUWEN PH (1992) Use of fluorescence induction to diagnose resistance of *Alopecurus myosuroides* Huds. (black-grass) to chlorotoluron. *Weed Research* **32,** 473–482.

VENCILL WK & FOY CL (1988) Distribution of triazine-resistant smooth pigweed (*Amaranthus hybridus*) and common lambsquarters (*Chenopodium album*) in Virginia. *Weed science* **36,** 497–499.

VOSS M, RENGER G, KÖTTER C & GRÄBER P (1984) Fluorometric detection of photosystem II herbicide penetration and detoxification in whole leaves. *Weed science* **32,** 675–680.

WANG P, PETEINATOS G, LI H et al. (2018) Rapid monitoring of herbicide-resistant *Alopecurus myosuroides* Huds. using chlorophyll fluorescence imaging technology. *Journal of Plant Diseases and Protection* **125,** 187–195.

WANG P, PETEINATOS G, LI H & GERHARDS R (2016) Rapid in-season detection of herbicide resistant *Alopecurus myosuroides* using a mobile fluorescence imaging sensor. *Crop Protection* **89,** 170–177.

WANG P, PETEINATOS GG & GERHARDS R (2017) In field identification of herbicide resistant *Apera spica-venti* using chlorophyll fluorescence. *Advances in Animal Biosciences: Precision Agriculture* **8,** 283–287.

WEBER JF, KUNZ C, PETEINATOS GG, SANTEL H-J & GERHARDS R (2017) Utilization of chlorophyll fluorescence imaging technology to detect plant injury by herbicides in sugar beet and soybean. *Weed technology* **31,** 523–535.

WILSON RG (1999) Response of nine sugarbeet (*Beta vulgaris*) cultivars to postemergence herbicide applications. *Weed technology* **13,** 25–29.

WILSON RG, YONTS CD & SMITH JA (2002) Influence of glyphosate and glufosinate on weed control and sugarbeet (*Beta vulgaris*) yield in herbicide-tolerant sugarbeet. *Weed technology* **16,** 66–73.

XUE J & SU B (2017) Significant remote sensing vegetation indices: a review of developments and applications. *Journal of Sensors*, doi.org/10.1155/2017/1353691.

ZHANG CJ, LIM SH, KIM JW, NAH G, FISCHER A & KIM DS (2016) Leaf chlorophyll fluorescence discriminates herbicide resistance in *Echinochloa* species. *Weed Research* **56,** 424–433.